Head Dress of Imperial Germany
1880-1916

Head Dress of Imperial Germany 1880-1916

Paul Sanders

Schiffer Military History
Atglen, PA

Preface

For over thirty years my family collected the helmets in this book. Just as much a part of the hobby, are the many fine memories of the fellow collectors and deals at the military shows and auctions. I have enjoyed the spirit of the chase for a special helmet, the meticulous study of the details of the helmets, and the regimental history research. Unfortunately, the only part I can share is the result. The helmets on the following pages are the result of a lifetime of enjoyable collecting. It is the intent of this book to share some of my enjoyment with all the new and old collectors of Imperial German helmets.

• • •

This book is a compilation of helmets collected over the past thirty years, and is intended as an illustrated reference source on the elaborate head dress of the Imperial German Army from 1880-1916. Displaying many rare helmets in full color, large format photographs, the book covers most branches within the army and is organized in alphabetic order beginning with the artillery units, moving through the various cavalry (hussars, dragoons, currasiers) and infantry units, and ending with uhlans – a superb cross section of all the different head dress types. Head dress of the large states like Prussia, Bavaria, Württemburg and Saxony are well represented, and in addition, smaller states like Hesse, Lubeck, Hamburg and the Saxon duchies can also be found. Other sections are devoted to the helmets of the generals and their staffs, with another showing six examples of the different ranks and models of the rare Garde du Korps helmets. This book is a beautiful, and highly detailed source for collectors.

Photography and book design by Robert Biondi.

Library of Congress Catalog Number: 2001086069.

Printed in China.
ISBN: 0-7643-1387-8

We are always looking for people to write books on new and related subjects. If you have an idea for a book, please contact us at the address below.

Published by Schiffer Publishing Ltd.
4880 Lower Valley Road
Atglen, PA 19310
Phone: (610) 593-1777
FAX: (610) 593-2002
E-mail: Schifferbk@aol.com.
Visit our web site at: www.schifferbooks.com
Please write for a free catalog.
This book may be purchased from the publisher.
Please include $3.95 postage.
Try your bookstore first.

In Europe, Schiffer books are distributed by:
Bushwood Books
6 Marksbury Ave.
Kew Gardens
Surrey TW9 4JF
England
Phone: 44 (0)208 392-8585
FAX: 44 (0)208 392-9876
E-mail: Bushwd@aol.com.
Free postage in the UK. Europe: air mail at cost.
Try your bookstore first.

Foreword

The fascination with and collection of Imperial German military headgear has its origins from the battlefields of France as early as 1914. Photographs from that time record victorious Tommies and Poilus posing with trophy pickelhauben taken from their enemies, but unfortunately those early victories were not as numerous as the Allies might have hoped. When the United States entered the fray, every American doughboy was on the lookout for a Luger and or a spiked or steel coalscuttle helmet to bring home as a souvenir. Many thousands of all three came to America after November 11, 1918. To add to this total, the United States government shipped back huge quantities of unissued Imperial German helmets which were given away as incentives to Liberty Bond purchasers during the many bond drives after the end of hostilities, and pictures of warehouses full of these treasures were published, making today's collector wonder where they all are today.

Another influx of Imperial helmets occurred after World War II when veterans returning from duty in Germany brought back thousands of the more exotic types and styles of these handsome helmets, many of which were manufactured originally as full dress peacetime examples for parade use.

The serious collecting of Imperial helmets in this country began in the 1950s and has continued unabated since that time. Many of the advanced collectors began to seriously study their passion, and as a result, discovered the vast variety and subtle differences of helmets and helmet plates. Books, both good and not so good, have been written in French, German, and English covering the subject in various depths of scholarly research, but with the advent of this particular volume, a more visually detailed and colorful coverage is brought to the modern collector and historian alike.

This coverage can give those interested a much clearer view of these rare and beautiful helmets than could be obtained from actually visiting European museums, as well as including and encompassing many examples of great rarity not to be commonly found elsewhere. The helmets seen in these pages portray the culmination of many long years of careful selection by a small group of discriminating American collectors and could not be found anywhere in a single location, or single volume.

The values of these stunning relics of a grandeur now past have risen dramatically over the past few years as they have become more difficult to find and as the economy of Germany has become robust and those collectors in that country are bringing their treasures back home.

Beauty is said to be in the eyes of the beholder and it would seem that anyone viewing the treasures so tastefully depicted on the following pages could not but help admire their design and execution in leather, steel, brass, frosted silver, gold wash and fine enameling. Ah, the ancient pride in being a soldier.

Charles Woolley
January 2001

Contents

• Prussian NCO Field Artillery (Feldartillerie)

• Prussian NCO Field Artillery profile (Feldartillerie)

• Prussian NCO 10th Field Artillery (Feldartillerie Regiment von Scharnhorst [1.Hannoversches Nr.10])

• Prussian NCO 10th Field Artillery profile (Feldartillerie Regiment von Scharnhorst [1.Hannoversches Nr.10])

• Hessian 25th Artillery Officer with parade bush (1916 model) (1. Grossherzoglich Hessisches Feldartillerie Regiment Nr.25)

• Hessian 25th Artillery Officer with parade bush profile (1. Grossherzoglich Hessisches Feldartillerie Regiment Nr.25)

• Hessian 25th Artillery Officer (1. Grossherzoglich Hessisches Feldartillerie Regiment Nr.25)

• Bavarian Artillery Officer with bush

• Prussian Artillery Officer 1915 War Model

• Prussian Artillery Officer 1915 War Model profile

• Oldenburg 19th Dragoon with parade bush (Oldenburgisches Dragoner Regiment Nr.19)

• Prussian 1st Guard Dragoon Officer (1.Garde Dragoner Regiment Königin Viktoria von Großbritannien und Irland)

• Prussian 2nd Guard Dragoon Officer (2.Garde Dragoner Regiment Kaiserin Alexandra von Russland)

• Prussian 2nd Guard Dragoon Officer profile (2.Garde Dragoner Regiment Kaiserin Alexandra von Russland)

• Prussian 1st Dragoon Officer with parade bush (Dragoner Regiment Prinz Albrecht von Preussen [Litthauisches] Nr.1)

• Prussian 1st Dragoon Officer with parade bush profile (Dragoner Regiment Prinz Albrecht von Preussen [Litthauisches] Nr.1)

• Prussian 9th Dragoon Officer (Dragoner Regiment König Carl I. von Rumänien [1.Hannoversches] Nr.9)

• Prussian 9th Dragoon Officer profile (Dragoner Regiment König Carl I. von Rumänien [1.Hannoversches] Nr.9)

• Prussian 16th NCO Dragoon (2.Hannoversches Dragoner Regiment Nr.16)

• Prussian 16th NCO Dragoon profile (2.Hannoversches Dragoner Regiment Nr.16)

• Oldenburg 19th Dragoon (Oldenburgisches Dragoner Regiment Nr.19)

• Oldenburg 19th Dragoon profile (Oldenburgisches Dragoner Regiment Nr.19)

• Hessian 23rd Dragoon Officer reserve (Garde Dragoner Regiment [1.Grossherzoglische Hesse] Nr.23)

• Hessian 23rd Dragoon Officer reserve profile (Garde Dragoner Regiment [1.Grossherzoglische Hesse] Nr.23)

• Hessian 23rd Dragoon Officer with 1916 star (Garde Dragoner Regiment [1.Grossherzoglische Hesse] Nr.23)

• Hessian 23rd Dragoon Officer with 1916 star profile (Garde Dragoner Regiment [1.Grossherzoglische Hesse] Nr.23)

• Baden Flugeladjudent with feather bush

• Bavarian General Staff with feather bush

• Prussian General with feather bush

• Saxon General with feather bush

• Saxon General Staff

• Saxon General Staff profile

• Württemberg General Staff with parade bush

• Württemberg General Staff with parade bush profile

• Württemberg General with feather bush

• Prussian Leib Garde Husaren Regiment Officer complete to parade

• Prussian 1st "Death Head" Hussar Officer complete to parade (1.Leib Husraen Regiment Nr.1)

• Prussian 2nd "Death Head" Hussar Officer complete to parade (2.Leib Husaren Regiment Königen Viktoria von Preussen Nr.2)

• Prussian 4th Hussar Officer reserve complete to parade (Husaren Regiment von Schill [1.Schlesisches] Nr.4)

• Prussian 7th Hussar Officer complete to parade (Husaren Regiment König Wilhelm I (1.Rheinisches) Nr.7)

• Prussian 9th Hussar NCO (2. Rheinisches Husaren Regiment Nr.9)

• Prussian 9th Hussar NCO (2. Rheinisches Husaren Regiment Nr.9)

• Prussian 9th Hussar Officer complete to parade (2. Rheinisches Husaren Regiment Nr.9)

• Prussian 11th Hussar Officer complete to parade (2.Westfalisches Husaren Regiment Nr.11)

• Prussian 12th Hussar Officer reserve complete to parade (Thuringisches Husaren Regiment Nr.12)

• Prussian 15th Hussar Officer complete to parade (Husaren Regiment Königen Wilhemina der Niederlande [Hannoversches] Nr.15)

• Brunswick 17th Hussar Officer complete to parade (Braunschweigisches Husaren Regiment Nr.17)

• Brunswick 17th Hussar Officer complete to parade profile (Braunschweigisches Husaren Regiment Nr.17)

• Brunswick 17th EM busby complete to parade (Braunschweigisches Husaren Regiment Nr.17)

• Saxon 18th Hussar Officer complete to parade (Early model) (Königlich Sächsisches 1.Husaren Regiment König Albert Nr.18)

• Saxon 20th Hussar Officer complete to parade (Late model) (Königlich Sächsisches 3.Husaren Regiment Nr.20)

• Prussian Kasier Alexander Garde Grenadier 1st Regiment Officer Miter (Kaiser Alexander der Garde Grenadier Regiment Nr.1 – Offizier-Grenadiermütze)

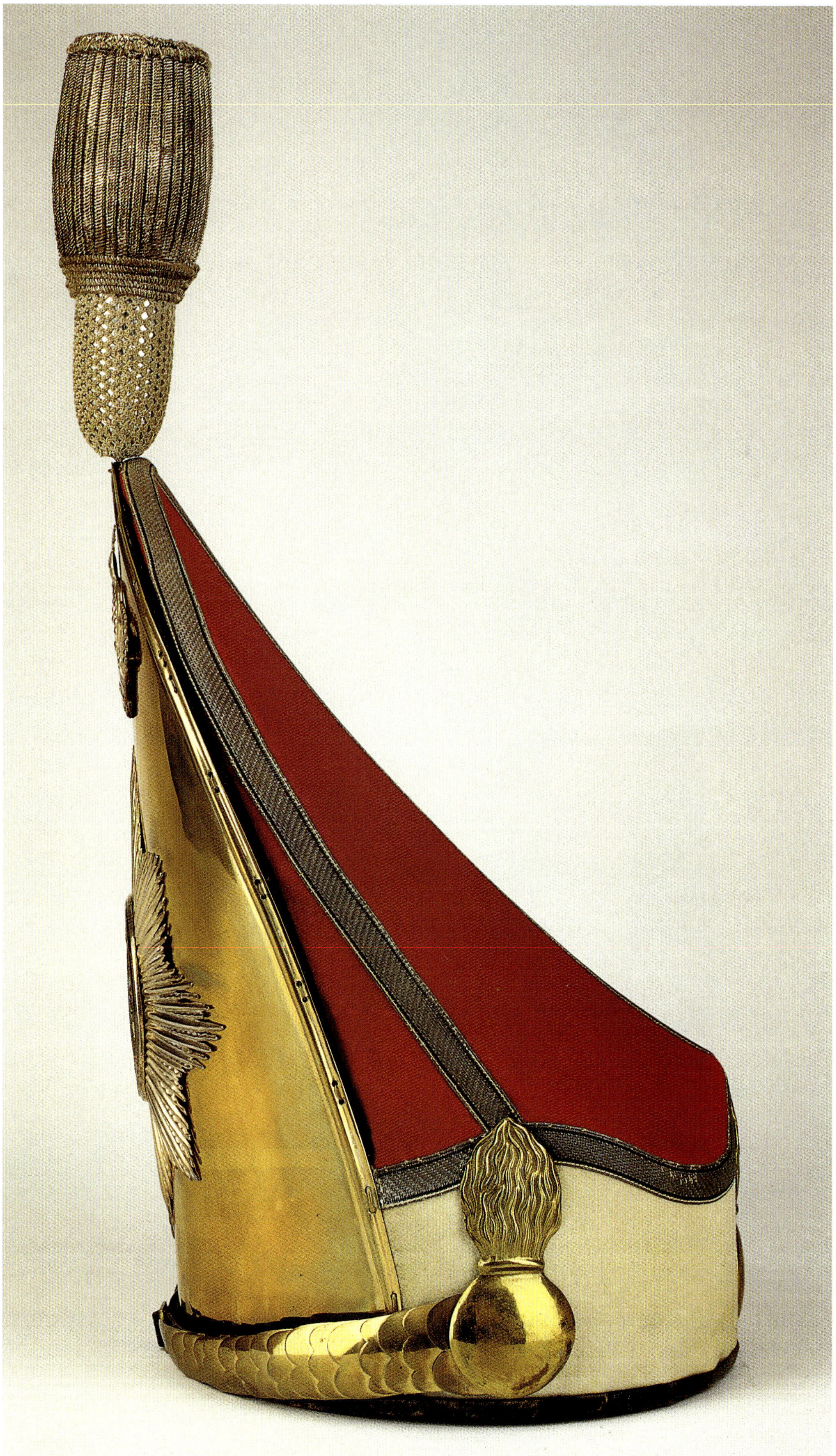

• Prussian Kasier Alexander Garde Grenadier 1st Regiment Officer Miter profile (Kaiser Alexander der Garde Grenadier Regiment Nr. 1 – Offizier-Grenadiermütze)

• Prussian Kasier Alexander Garde Grenadier 1st Regiment Officer Miter back (Kaiser Alexander der Garde Grenadier Regiment Nr. 1 – Offizier-Grenadiermütze)

• Prussian 1st Garde zu Fuss Officer Miter (1.Garde Regiment zur Fuß Offizier Grenadiermütze)

• Prussian 1st Garde zu Fuss Officer Miter profile and back (1.Garde Regiment zur Fuß Offizier Grenadiermütze)

• Prussian 1st Grenadier Officer (Grenadier Regiment Kronprinz [1.Ostpreussisches] Nr.1)

• Prussian 1st Grenadier Officer (Grenadier Regiment Kronprinz [1.Ostpreussisches] Nr.1)

• Prussian Line Grenadier Officer (1915 model)

• Prussian Line Grenadier Officer (1915 Model) profile

• Prussian 8th Grenadier Officer (1.Brandenburgisches Leib-Grenadier Regiment Nr.8)

• Prussian 8th Grenadier Officer profile (1.Brandenburgisches Leib-Grenadier Regiment Nr.8)

• Prussian 74th Infantry Officer reserve (1.Hannoversches Infanterie Regiment Nr.74)

• Prussian 74th Infantry Officer reserve profile (1.Hannoversches Infanterie Regiment Nr.74)

• Prussian 76th Hamburg Infantry Officer (2.Hannoversches Infanterie Regiment Nr.76)

• Prussian 76th Hamburg Infantry Officer profile (2.Hannoversches Infanterie Regiment Nr.76)

• Prussian 83rd Waldeck-Pyrmont Infantry III Bataillon NCO (Infanterie Regiment von Wittich [3.Kurhessisches] Nr.83)

• Prussian 83rd Waldeck-Pyrmont Infantry III Bataillon NCO profile (Infanterie Regiment von Wittich [3.Kurhessisches] Nr.83)

• Prussian 87th Nassauisches Infantry Officer (1.Nassauisches Infanterie Regiment Nr.87)

• Prussian 87th Nassauisches Infantry Officer profile (1.Nassauisches Infanterie Regiment Nr.87)

• Prussian 88th Nassauisches Line Officer (2.Nassauisches Infanterie Regiment Nr.88)

• Prussian 88th Nassauisches Line Officer profile (2.Nassauisches Infanterie Regiment Nr.88)

• Mecklenburg-Schwerin 89th I and II Bataillon Infantry Officer Reserve (Grossherzoglich Mecklenburgisches Grenadier Regiment Nr.89)

• Mecklenburg-Schwerin 89th I and III Bataillon Infantry Officer Reserve profile (Grossherzoglich Mecklenburgisches Grenadier Regiment Nr.89)

• Oldenburg 91st Infantry Officer (Oldenburgisches Infanterie Regiment Nr.91)

• Oldenburg 91st Infantry Officer profile (Oldenburgisches Infanterie Regiment Nr.91)

• Anhalt 93rd Infantry Officer Reserve (Anhaltisches Infanterie Regiment Nr.93)

• Anhalt 93rd Infantry Officer Reserve profile (Anhaltisches Infanterie Regiment Nr.93)

• Saxon-Weimar 94th Thüringisches Infantry Officer reserve (Infanterie Regiment Grossherzoglich von Sachsen [5.Thüringinisches] Nr.94)

• Saxon-Weimar 94th Thüringisches Infantry Officer reserve profile (Infanterie Regiment Grossherzoglich von Sachsen [5.Thüringinisches] Nr.94)

• Saxon-Meiningen-Coburg-Gotha 95th Thüringisches Infantry Officer (6.Thüringinisches Infanterie Regiment Nr.95)

• Saxon-Meiningen-Coburg-Gotha 95th Thüringisches Infantry Officer profile (6.Thüringinisches Infanterie Regiment Nr.95)

• Schwarzburg-Rudolstadt 96th III Bataillon Thüringisches Infantry Officer reserve (7.Thüringinisches Infanterie Regiment Nr.96)

• Schwarzburg-Rudolstadt 96th III Bataillon Thüringisches Infantry Officer reserve profile (7.Thüringinisches Infanterie Regiment Nr.96)

• Reuss 96th II Bataillon Thüringisches Infantry Officer (7.Thüringinisches Infanterie Regiment Nr.96)

• Reuss 96th II bataillon Thüringisches Infantry Officer profile (7.Thüringinisches Infanterie Regiment Nr.96)

• Prussian 98th Infantry EM (Metzer Infanterie Regiment Nr.98)

• Prussian 98th Infantry EM profile (Metzer Infanterie Regiment Nr.98)

• Saxon 100th Leib Grenadier Infantry Officer (Königlich Sächsisches 1.Leib-Grenadier Regiment Nr.100)

• Saxon 100th Leib Grenadier Infantry Officer profile (Königlich Sächsisches 1.Leib-Grenadier Regiment Nr.100)

• Baden 109th Leib Grenadier Infantry Officer (1.Badisches Leib-Grenadier Regiment Nr.109)

• Baden 109th Leib Grenadier Infantry Officer profile (1.Badisches Leib-Grenadier Regiment Nr.109)

• Baden 109th Leib Grenadier Infantry Officer reserve (1.Badisches Leib-Grenadier Regiment Nr.109)

• Baden 109th Leib Grenadier Infantry Officer reserve profile (1.Badisches Leib-Grenadier Regiment Nr.109)

• Hessian 115th Leib Garde Infantry Officer (Leib Garde Infanterie Regiment [1.Grossherzoglich Hessische] Nr.115)

• Hessian 115th Leib Garde Infantry Officer profile (Leib Garde Infanterie Regiment [1.Grossherzoglich Hessische] Nr.115)

• Hessian 115th Leib Garde Infantry Officer reserve with parade bush (Leib Garde Infanterie Regiment [1.Grossherzoglich Hessische] Nr.115)

• Hessian 115th Leib Garde Infantry Officer reserve with parade bush profile (Leib Garde Infanterie Regiment [1.Grossherzoglich Hessische] Nr.115)

• Hessian 117th Infantry Officer (Infanterie Leibregiment Grossherzogin [3.Grossherzolich Hessisches] Nr.117)

• Hessian 117th Infantry Officer profile (Infanterie Leibregiment Grossherzogin [3.Grossherzolich Hessisches] Nr.117)

• Hessian 118th Infantry Officer reserve (Infanterie Regiment Prinz Carl [4.Grossherzoglich Hessisches] Nr.118)

• Hessian 118th Infantry Officer reserve profile (Infanterie Regiment Prinz Carl [4.Grossherzoglich Hessisches] Nr.118)

• Prussian 162nd Lubeck Infantry Officer (Infanterie Regiment Lübeck [3.Hanseatisches] Nr.162)

• Prussian 162nd Lubeck Infantry Officer profile (Infanterie Regiment Lübeck [3.Hanseatisches] Nr.162)

• Bavarian Infantry Officer

• Bavarian Infantry Officer profile

• Prussian Garde Infantry EM

• Prussian Garde Infantry EM profile

• Prussian Garde Infantry NCO

• Prussian Garde Infantry NCO Profile

• Prussian Landwehr Infantry Officer

• Prussian Landwehr Infantry Officer profile

• Prussian Infantry EM War Model

• Prussian Infantry EM War Model profile

• Prussian Guard Jäger Bataillon Officer with bush

• Prussian Guard Jäger Bataillon Officer with bush profile

• Prussian 7th Jäger Bataillon Schaumburg Lippe NCO (Westfälisches Jäger Bataillon Nr.7)

• Prussian 7th Jäger Bataillon Schaumburg Lippe NCO profile (Westfälisches Jäger Bataillon Nr.7)

• Prussian 10th Jäger Bataillon Officer (Hannoversches Jäger Bataillon Nr.10)

• Prussian 10th Jäger Bataillon Officer profile (Hannoversches Jäger Bataillon Nr.10)

• Reiten Des Feld Jäger Officer with bush

• Reiten Des Feld Jäger Officer with bush profile

• Prussian 1st Jäger zu Pferde Officer (Regiment Königs Jäger zu Pferde Nr.1)

• Prussian 1st Jäger zu Pferde Officer profile (Regiment Königs Jäger zu Pferde Nr.1)

• Prussian 7th Jäger zu Pferde EM (Jäger Regiment zu Pferde Nr.7)

• Prussian 7th Jäger zu Pferde EM profile (Jäger Regiment zu Pferde Nr.7)

• Prussian 8th Jäger zu Pferde Officer (Jäger Regiment zu Pferde Nr.8)

• Prussian 8th Jäger zu Pferde Officer profile (Jäger Regiment zu Pferde Nr.8)

• Prussian Garde du Corps EM with parade eagle (Regiment der Garde du Corps)

• Prussian Garde du Corps EM with parade eagle profile (Regiment der Garde du Corps)

• Prussian Garde du Corps NCO with parade eagle (Regiment der Garde du Corps)

• Prussian Garde du Corps NCO with parade eagle profile (Regiment der Garde du Corps)

• Prussian Garde du Corps Cadet/Fähnrich with parade eagle (Regiment der Garde du Corps)

• Prussian Garde du Corps Cadet/Fähnrich with parade eagle profile (Regiment der Garde du Corps)

• Prussian Garde du Corps Officer with parade eagle (early Model) (Regiment der Garde du Corps)

• Prussian Garde du Corps Officer with parade eagle profile (early Model) (Regiment der Garde du Corps)

• Prussian Garde du Corps Officer (late model) (Regiment der Garde du Corps)

• Prussian Garde du Corps Officer (late model) profile (Regiment der Garde du Corps)

• Prussian 1st Leib Kurassier Officer Reserve (Leib Kürassier Regiment Grosser Kurfürst [Schlesisches] Nr.1)

• Prussian 1st Leib Kurassier Officer Reserve profile (Leib Kürassier Regiment Grosser Kurfürst [Schlesisches] Nr.1)

• Prussian 2nd Kurassier Officer (Kürassier Regiment Königin [Pommersches] Nr.2)

• Prussian 2nd Kurassier Officer profile (Kürassier Regiment Königin [Pommersches] Nr.2)

• Prussian Line Kurassier Officer

• Prussian Line Kurassier Officer profile

• Prussian Line Kurassier NCO

• Prussian Line Kurassier NCO profile

• Prussian 6th Kurassier Officer (Kürassier Regiment Kaiser Nikolaus I von Russland [Brandenburgisches] Nr.6)

• Prussian 6th Kurassier Officer profile (Kürassier Regiment Kaiser Nikolaus I von Russland [Brandenburgisches] Nr.6)

• Prussian Line Kurassier EM

• Prussian Line Kurassier EM profile

• Saxon Guard Reiter Officer reserve with parade lion (Königlich Sächsisches Garde Reiter Regiment [1.Schweres Regiment])

• Saxon Garde Reiter Officer reserve with parade lion profile (Königlich Sächsisches Garde Reiter Regiment [1.Schweres Regiment])

• Saxon Garde Reiter EM (Königlich Sächsisches Garde Reiter Regiment [1.Schweres Regiment])

• Saxon Garde Reiter EM profile (Königlich Sächsisches Garde Reiter Regiment [1.Schweres Regiment])

• Saxon Karabiner Officer with bush (Königlich Sächsisches Karabinier Regiment [2.Schweres Regiment])

• Saxon Karabiner Officer with bush profile (Königlich Sächsisches Karabinier Regiment [2.Schweres Regiment])

• Prussian Line Machine Gun Abteilung

• Prussian Line Machine Gun Abteilung profile

• Prussian Marine Seebataillon Officer

• Prussian Marine Seebataillon Officer profile

• Prussian Guard Pionier Officer (Garde Pionier Bataillon)

• Prussian Guard Pionier Officer profile (Garde Pionier Bataillon)

• Prussian 10th Pionier Officer (Hannoversches Pionier Bataillon Nr.10)

• Prussian 10th Pionier Officer profile (Hannoversches Pionier Bataillon Nr.10)

• Prussian Pionier Officer (1915 model)

• Prussian Pionier Officer (1915 model) profile

• Prussian Telegraph Officer (Telegraphen Bataillon)

• Prussian Telegraph Officer profile (Telegraphen Bataillon)

• Prussian Garde Train Officer (Garde Train Bataillon)

• Prussian Garde Train Officer profile (Garde Train Bataillon)

• Prussian Eisenbahn EM (Eisenbahn Regiment Nr.1-3)

• Prussian Eisenbahn EM profile (Eisenbahn Regiment Nr.1-3)

• Prussian 1st Garde Uhlan Officer complete to parade (1.Garde Ulanen Regiment)

• Prussian 3rd Garde Uhlan Officer complete to parade (3.Garde Ulanen Regiment)

• Prussian 1st Uhlan Officer complete to parade (Ulanen Regiment Kaiser Alexander III. von Russland [Westpreussisches] Nr.1)

• Prussian 2nd Uhlan Officer complete to parade (Ulanen Regiment von Katzler [Schlesisches] Nr.2)

• Prussian 3rd Uhlan Reserve Officer complete to parade (Ulanen Regiment Kaiser Alexander II. von Russland [1.Brandenburgisches] Nr.3)

• Prussian 7th Uhlan Officer complete to parade (Ulanen Regiment Grossherzog Friedrich von Baden [Rheinisches] Nr.7)

• Prussian 9th Uhlan Officer complete to parade (2.Pommersches Ulanen Regiment Nr.9)

• Prussian 13th Uhlan Dienst Officer (Königs Ulanen Regiment [1.Hannoversches] Nr.13)

• Prussian 13th Uhlan Officer reserve complete to parade (Königs Ulanen Regiment [1.Hannoversches] Nr.13)

• Prussian 13th Uhlan Officer War Model with cords (Königs Ulanen Regiment [1.Hannoversches] Nr.13)

• Prussian 14th Uhlan Officer complete to parade (2.Hannoversches Ulanen Regiment Nr.14)

• Saxon 17th Uhlan complete to parade (1.Ulanen Regiment Nr.17 Kaiser Franz Josef von Osterreich, König von Ungaren)

• Saxon 18th Uhlan Officer reserve complete to parade (Königlich Sächsisches 2.Ulanen Regiment Nr.18)

• Saxon 18th Uhlan Officer complete to parade (Model 1916) (Königlich Sächsisches 2.Ulanen Regiment Nr.18)

• Württemberg 19th Uhlan Officer complete to parade (Ulan Regiment König Karl [1.Württemburgisches] Nr.19)

• Württemberg 20th Uhlan Officer complete to parade (Ulanen Regiment König Wilhelm I [2.Württembergische] Nr.20)

• Saxon 21st Uhlan Officer complete to parade (Königlich Sächsisches 3.Ulanen Regiment Nr.21 Kaiser Wilhelm König von Preussen)

• Saxon 21st Uhlan Officer Dienst

• Bavarian 1st Uhlan Officer complete to parade (1.Ulanen Regiment Kaiser Wilhelm II. König von Preussen)

• Bavarian 2nd Ulhlan Reserve Officer complete to parade (2.Ulanen Regiment König)

• Above: White silk lining of Officer Infantry helmet. Below: Brown silk lining of Officer Jäger helmet

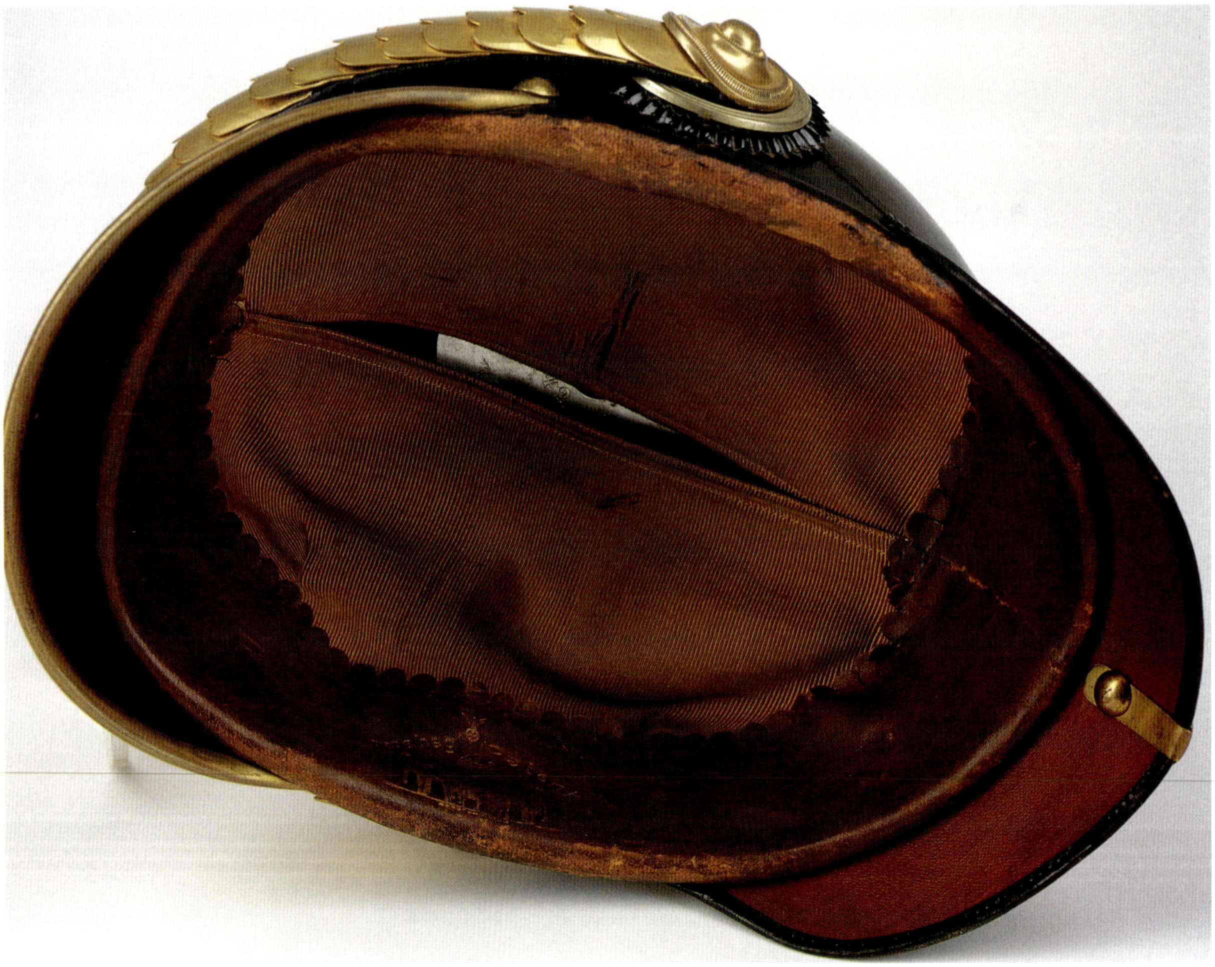